Fruit Bouquets

For Celebrations!

Create Your Own Gifts & Centerpieces

Delicious Designs

Printed in the United States of America
by G&R Publishing Co.

Published By:

507 Industrial Street
Waverly, IA 50677

ISBN-13: 978-1-56383-341-0
ISBN-10: 1-56383-341-7
Item #3624

Table of Contents

Getting Started

Creating a fruit bouquet can make celebrating an important event or day even more special. Whether used as an edible party centerpiece or given as a unique homemade gift, a fresh fruit arrangement is sure to please and impress.

Practice Safety

Wash your hands and utensils thoroughly with soap and water before handling fresh fruits and vegetables. Make sure your work surface is clean and sanitary as well. Use well-sharpened knives properly with a cutting board underneath and practice general kitchen safety when handling sharp utensils.

Choose the freshest vegetables and fruits available. Wash them thoroughly under running water to remove dirt and bacteria. Prior to cutting or peeling fruits and vegetables such as pineapples, melons or carrots, scrub

the hard outer rind or firm skin under running water with a vegetable brush. Drain and/or pat produce dry with paper towels. For best results, soak clean, sliced vegetables in cold water or ice water for 15 minutes or more before attaching to bouquet base.

Waxes are often applied to produce such as apples, cucumber and zucchini, to help retain moisture. Therefore, do not wash these fruits and vegetables until you are ready to create your bouquets in order to keep them firm and crisp as long as possible.

It is important to keep most fruits and vegetables at a cool temperature while preparing and arranging your bouquet. *(Exception: Watermelon should be at room temperature for carving; clean cuts will be easier to make when the watermelon is not chilled.)* After pieces of the bouquet are cut, refrigerate them if they are not going to be immediately assembled. For optimal freshness and beauty, serve fruit bouquets just after assembly. If holding time is required, loosely cover bouquet

continued

and refrigerate. If transporting, pack and secure the loosely covered bouquet in a large cooler. If given as a gift, encourage the recipient to enjoy it promptly and refrigerate leftovers.

Prepare the Base

Selecting the perfect container for your fruit bouquet is part of the celebration fun; choose one that communicates the theme and adds interest. A sturdy flat base and a wide-mouth opening are important. Wash and thoroughly dry containers before filling.

A head of iceberg lettuce placed inside the container is a great, inexpensive way to secure the fruit flowers or other skewers in bouquets. It can be easily cut to fit inside smaller containers, or a whole head and/or pieces can fill larger ones. Heads of cabbage suit some arrangements, such as Star Spangled Berries on page 24, but the surface will be a little more difficult to puncture.

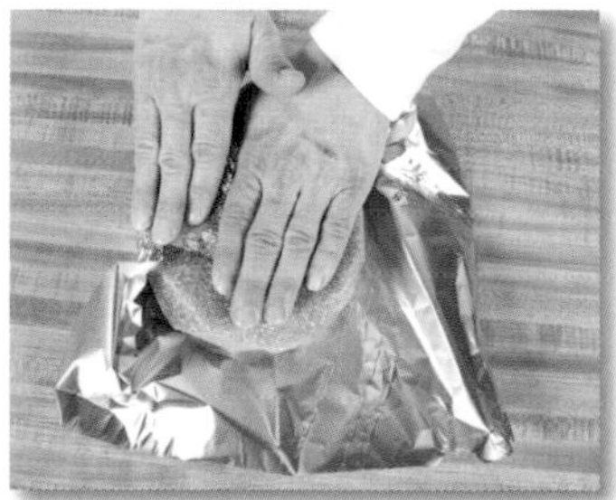

Styrofoam is used in the creation of other fruit bouquets. It may form the shape for the base, such as Sweet Heart on page 12, or make an arrangement more sturdy, such as Caramel Apple Gourmet Bouquet on page 48. Styrofoam should not be ingested. Cover it properly and do not allow foam dust to contaminate fruit. Wrap Styrofoam with aluminum foil before using near fruit. Discard any food in direct contact with Styrofoam. Styrofoam cones and other bases may be fastened to a serving platter with doublestick tape if necessary.

Finally, consider the use of non-traditional items for bouquet bases, such as a carved out whole fruit like a watermelon, pumpkin or pineapple. Or, use a specialty container such as a toy wagon or plastic beach pail. Don't be afraid to vary from the suggested containers featured on these pages, as many bouquets adapt well to different containers. For example, the cover bouquet can be arranged in a more casual base as shown.

Place the Fruit

In most cases, fruit is skewered to place it in the prepared base. A variety of items can be used for stem skewers, such as decorative plastic drink stirrers, plastic lollipop stems, wood craft sticks, cellophane frill toothpicks or round toothpicks. (Flat picks tend to break more easily.)

Frequently, 10″ to 12″ wooden or bamboo skewers are used, due to their low cost and easy availability. To trim wooden or bamboo skewers, snip to desired lengths with clean, sanitary pruning shears. These skewers can also be grouped and wrapped together to make a single, stronger stem. Use a wooden dowel if an extremely thick stem is needed for a heavy bloom or topiary.

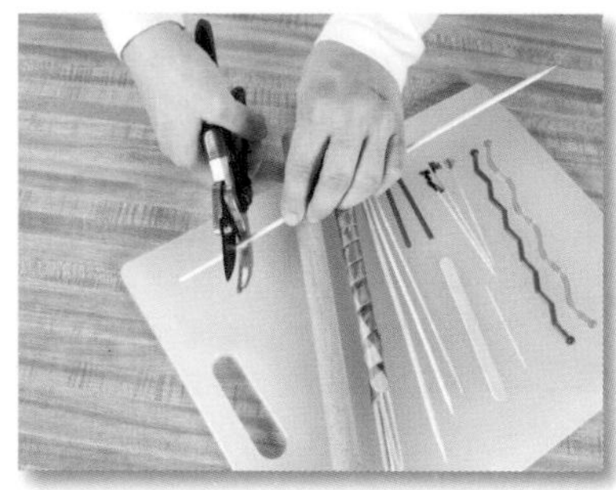

As an arranged bouquet of fruit sits, the fruit pieces may begin to slide down the stems. To prevent sliding, place a "stopper" on the skewer before adding the fruit. Wrap small craft rubber bands around each skewer below the fruit, or if you prefer to use a food item, slide a raisin or gumdrop onto the skewer.

Wrapping It Up

Wrap a beautiful fruit or vegetable bouquet in a large sheet of cellophane, gathering the corners up at the top of the arrangement and tucking it in at the sides. Fasten at the top with a pretty bow before presenting it as an incredible celebration gift.

Pieces of equipment

- Skewers/sticks/picks/ etc. (See previous section.)
- Round toothpicks
- Pruning shears
- Food-safe containers
- Ribbons or other embellishments
- Low-temperature hot glue gun
- Styrofoam ("foam")
- Florist's wire
- Knives, such as chef's, paring, carving, etc.
- Cutting boards
- Scissors
- Ripple potato slicer
- Melon baller
- Vegetable peeler
- Metal cookie cutters
- Sheet pans
- Waxed paper, plastic wrap and aluminum foil
- Zippered plastic bags
- Plastic piping bags fitted with decorative tips
- Gel or paste food coloring
- Sheet of Styrofoam covered with waxed paper

Sweet & Single

You will need:

- 1 head cabbage or iceberg lettuce
- Container (Sample uses a 4½″ square pot, 5½″ deep.)
- 6 (12″) wooden or bamboo skewers
- 1½ yds. decorative ribbon (Sample uses wired ribbon, 1½″ wide.)
- Low-temperature hot glue gun
- 1 small bunch green kale or leafy lettuce
- ¼ whole fresh pineapple with leaves
- 1 large bunch red, purple or black seedless grapes
- 12 green seedless grapes
- Toothpicks

To Begin...

1 Prepare the container. Trim sides and bottom of cabbage head as needed to fit snugly into container. Place the cabbage head in the container so the top rests 1″ below rim. Arrange kale leaves over cabbage but leave center of cabbage head exposed. With a skewer, poke a starter hole, 1½″ to 2″ deep, in the center of cabbage; enlarge it slightly. Set container aside.

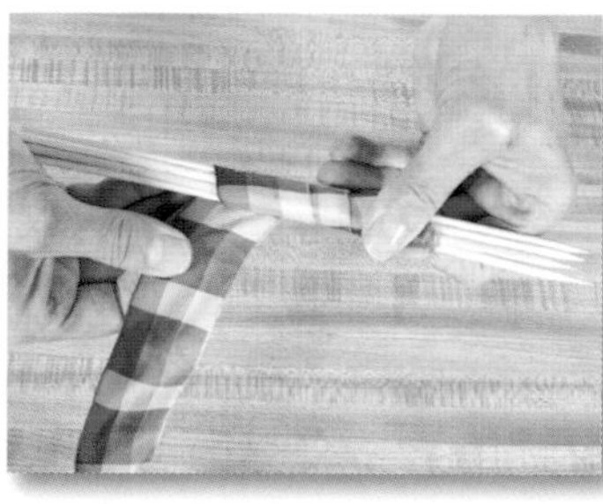

2 Cut a piece of ribbon 18″ long. Gather the 6 skewers together with points even. Starting 2″ from points, wrap ribbon diagonally around the skewers, pulling ribbon snugly to hold skewers together and make one thick stem. Leave the last 2″ of skewers unwrapped. Fasten ends of ribbon to skewers with hot glue.

3 Cut off the top of the pineapple. Pull out 2 leaves and reserve them for garnishing the finished bouquet. With a sharp knife, cut off pineapple skin. Place

continued

pineapple on its side and cut a crosswise slice, 1¼″ thick. Trim edges of slice to make a smooth round disk.

4 Insert points of wrapped skewer stem into the edge of the pineapple disk until points pierce the hard center core. Cover and chill until assembly.

5 To make the grape spears, thread red grapes onto toothpicks. Make approximately 10 spears with 1 grape, 20 spears with 2 grapes and 10 spears with 3 grapes. To lengthen the sticks on the 3-grape spears, insert a second toothpick in the opposite direction, allowing the tip of toothpick to stick out 1″. Avoid poking the toothpick through the top grape on each skewer.

6 Cut green grapes in half crosswise. Reserve the smooth halves for this bouquet; discard or use the stem-end halves for another use. Break 8 toothpicks in half; set aside.

7 With pineapple disk on a flat surface, insert prepared red grape spears around one edge of disk. Point the spears upward slightly. Start with a 3-grape spear inserted at the center top. Work in both directions around the pineapple to create a balanced bloom. Alternate spear lengths in the following pattern as shown: 3, 2, 1, 2, 3, etc. Leave an open space near stem.

8 Carefully turn disk over and place it on a small bowl or plastic container to work on back side of "flower." Repeat process to place grapes spears around back edge of pineapple disk.

9 Insert a toothpick half into the center of pineapple disk. Attach a red grape. Insert 6 toothpick halves around red grape as shown and attach the green grape halves on tips to surround the center grape. Repeat on other side of flower after assembly.

10 To assemble arrangement, insert skewer stem into the starter hole in cabbage, pressing down firmly until fruit "flower" stands straight and secure. (If necessary, remove pineapple flower head to push down on stem more easily; then replace flower head.) Insert reserved pineapple leaves in kale near stem to resemble flower's leaves. Tie remaining decorative ribbon around base as desired. Discard kale after fruit blossom has been eaten.

Celebration Suggestions...

Birthdays, Mother's Day, Garden Party, Congratulations, Get Well or say, "Cheer up!"

Sweet Heart

You will need:

- 2 (12 x 18″) sheets of paper
- 2 (12 x 18 x 2″) sheets Styrofoam
- Toothpicks
- Low-temperature hot glue gun
- 2 bunches green kale or leafy lettuce
- ½ small seedless watermelon
- ½ honeydew melon, seeded
- ½ fresh pineapple, cored
- ½ cantaloupe, seeded
- Melon baller
- 25 to 30 large strawberries
- ½ pt. fresh blueberries
- 1 large bunch red seedless grapes (about 200)
- 1 small bunch green seedless grapes

To Begin...

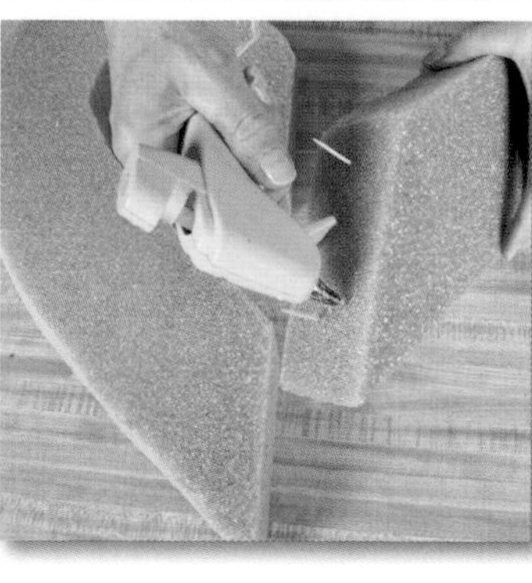

1 To create a heart pattern, tape the sheets of paper together along 18″ edges. Fold the paper in half at tape and sketch half a large heart shape to fill the space. When you like the shape, cut it out. Measure and draw a line 4″ from cut edge to create a smaller heart inside. Cut along line and remove small heart. Set the sheets of foam together with 18″ edges touching. Open up pattern heart and place it on top of foam, lining up seams of pattern with center edges of foam. Trace around heart pattern with a pencil. With a sharp knife, cut along tracings. Insert several toothpicks into center seams of one foam piece, apply glue and press the foam pieces together to join the seams, making one large, complete heart.

2 Cover foam heart with aluminum foil. Arrange pieces of leafy lettuce on top of foil, draping it over edges; use toothpicks to hold lettuce in place.

---

3 Cut watermelon and honeydew melon into thick slices; remove the rinds. Cut off the pineapple skin. Slice the fruits into cubes, 1″ to 1½″ in size.

---

4 With a melon baller, cut cantaloupe into large balls.

5 Use toothpicks to fasten fruit pieces to top and sides of heart base. First, attach a row of strawberries around center top of heart as shown, leaving space between them for blueberries.

6 Attach red grapes next. Push toothpicks into foam around lower outer edge of heart. (Push them through the lettuce far enough so only a short point sticks out.) Slide 1 grape onto each point, without piercing the top of the grape. Adjust toothpicks as needed or break toothpicks in half for smaller grapes. Attach a second row of grapes above first row around outside of heart in the same manner. Attach 1 to 2 rows of red grapes to the inside edge of heart.

7 Place a toothpick between pairs of strawberries on top and slide 2 or 3 blueberries on each toothpick. With toothpicks, attach watermelon, pineapple and honeydew cubes around the berries. Start with the largest pieces first and stagger fruits so color is well balanced and fruits are evenly spaced. For easiest assembly, push the toothpicks into the foam first and then press the pieces of fruit onto the point of toothpick. Fill in remaining exposed areas of heart with cantaloupe balls and green grapes. These fruits don't need to be pushed all the way down to the lettuce.

---

8 If desired, cover a large sturdy piece of cardboard with foil or parchment paper and set the fruit-covered heart centerpiece on top to transport or serve. Discard leafy lettuce after fruits have been eaten.

Tips:

- Purchase pre-cut fruit cubes and melon balls for faster assembly.
- Cut a smaller Styrofoam heart to serve fewer people, or purchase a small ready-to-use foam heart in a craft or floral department.

Celebration Suggestions...

Anniversary Party, Wedding, Bridal Shower, Open House, Housewarming Party, Valentine's Day or say, "I love you!"

Healthy Birthday To You

You will need:

- 1 watermelon (approx. 26″ circumference)
- Toothpicks
- 9″ to 10″ platter or cake plate
- 1 honeydew melon (approx. 20″ circumference)
- 1 cantaloupe (approx. 20″ circumference)
- 1 small bunch large red seedless grapes
- 1 small bunch large green seedless grapes
- ½ pt. blueberries
- 3 to 7 pineapple tidbits, drained well

Getting Started...

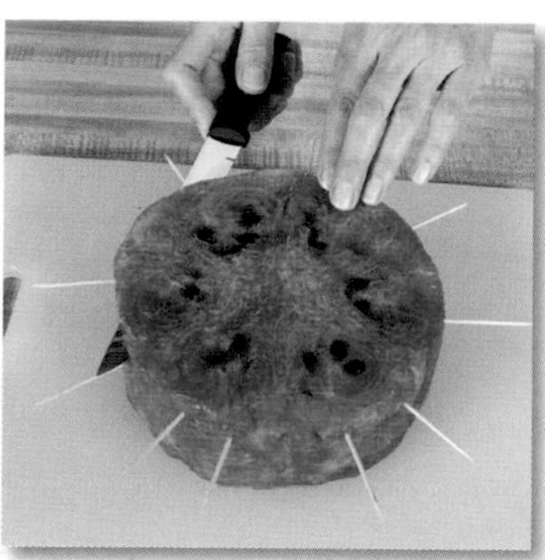

1 With a large sharp knife, cut a 3″ crosswise slice out of the center (largest) section of the watermelon. Trim off the rind and white flesh, leaving a round disk of pink flesh. Trim outside edges until disk is perfectly round. Place disk on a cutting board and carefully trim top and bottom as needed until watermelon disk is level and 2¼″ to 2½″ in height. (If desired, use toothpicks to mark cutting line as explained in "Cutting & Carving Tips" on page 58.) Remove any visible seeds. Transfer watermelon disk to the serving plate.

2 In the same manner, cut a 3″ wide slice across the center (largest) section of the honeydew melon to make a wide melon ring. Remove seeds and place honeydew ring on a cutting board. Cut off the rind and trim green flesh vertically until outside edge of disk is round and sides are straight up and down. Trim top and/or bottom until melon disk is level and about 2″ in height, using toothpicks as markers as needed. Set honeydew layer on top of watermelon layer to check fit. Honeydew diameter should be smaller than watermelon, leaving a ledge approximately 1″ wide between layers.

3 To make the cantaloupe layer, cut off 1″ from one end of the whole melon. Then cut a crosswise slice, 3″ from first cut. A thin layer of flesh should remain on narrowest cut end, making a cup shape; this will be the top of the cake. Carefully remove seeds. Cut off the rind. Trim orange flesh vertically to make the disk round, making sure cantaloupe piece remains larger than the center hole in the honeydew melon. Trim open end of melon "cup" to make it level and 1¾″ to 2″ in height.

4 To make grape flowers, slice a small piece off the stem end of several red and green grapes. To make 6 "petals", place grape on a cutting board, with the cut end up. Use a sharp knife to make 3 criss-cross cuts, going about ¾ of the way into grape toward the bottom, but not through it. Gently pry the petals apart with fingers and place a blueberry in the middle. Push a toothpick through the bottom of grape and into blueberry; set aside. Make 5 grape flowers.

5 To make each blueberry "candle", thread 4 to 6 blueberries on a toothpick; top with a pineapple tidbit. Set aside.

---

6 Cut 25 red grapes in half, lengthwise; set aside. Break toothpicks in half and poke pointed end into approximately 12 whole red grapes and 12 whole green grapes; set aside.

7 To assemble the cake, set honeydew layer on top of watermelon, inserting toothpicks between layers to prevent slippage, if necessary. Set cantaloupe layer on top, using toothpicks as needed. Gently push whole grape picks into the watermelon along the base, alternating colors. Insert toothpick with one grape flower on this layer. Set the cut edge of red grape halves around the honeydew and cantaloupe layers, stem ends down, as shown. Insert the toothpick end of three grape flowers into honeydew layer and remaining grape flower on cantaloupe layer. Insert blueberry "candles" around top edge of cantaloupe layer.

Tips:

- The center of the cantaloupe top is quite thin so if you wish to place a candle there, use a thin 10" bamboo skewer that will reach down into the watermelon layer for support.
- For economy, open one snack-size container of fruit cocktail or pineapple tidbits for use on the blueberry "candles."
- If the honeydew melon is large, it can be cut like the cantaloupe so the layer has a cup shape rather than a ring shape.
- Each cake layer can be wider or narrower according to personal preferences. Sample shows layers getting progressively narrower from bottom to top.

Variations:

- *Omit the pineapple tidbits and use 4" yellow cellophane frill toothpicks for the "flame" on the blueberry "candles."*
- *When in season, place thin slices of star fruit under blueberry candles or use them in place of grape garnishes.*

Celebration Suggestions...

Happy Birthday, or omit the candles and top the cake with extra grape flowers for a Surprise Party, Retirement Party, Anniversary, Bridal Shower or Summer Wedding.

Personal Crazy Daisy

Makes 6 individual place-setting arrangements

You will need:

- 1 large red apple
- Lemon juice or anti-browning product for fruit
- 1 whole fresh pineapple
- 3″ flower-shaped metal cookie cutter
- ½ cantaloupe, seeded
- ½ honeydew melon, seeded
- 6 glass votive cups, trimmed with ribbon or rick-rack (Sample uses clear glass flower pots, 2½″ deep.)
- 6 (6″) green plastic lollipop sticks
- Toothpicks

To Begin...

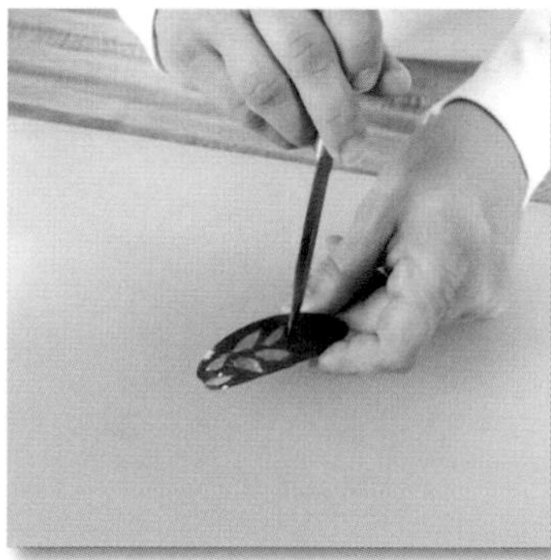

1 Core and cut the apple into 6 wedges. Using a paring knife, create more of a "leaf" shape by carving a little from the sides of each wedge, near the corners. Use a small sharp knife to carefully carve the "veins" of a leaf into the skin of each wedge. Carefully pop out the carved pieces using the knife tip, leaving the fruit's flesh exposed against the contrasting skin. Soak the apples in a mixture of water and lemon juice or a commercially prepared anti-browning product and then chill. (See page 58 for complete instructions.)

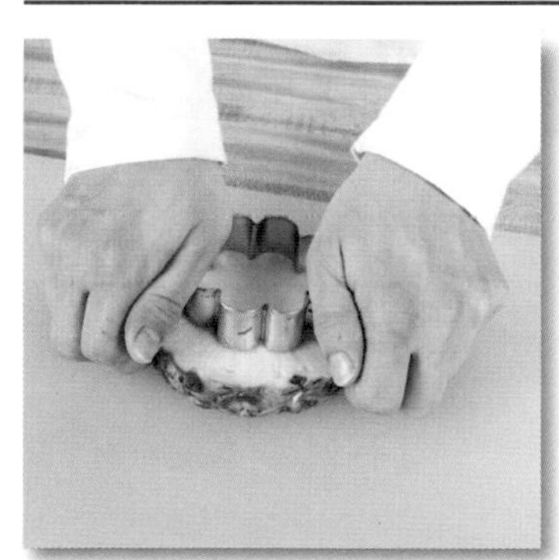

2 Slice the pineapple into 6 (¾″) disks, making one for each flower pot. To cut a daisy flower, center the cookie cutter over a pineapple disk. (Metal cookie cutters are recommended for a clean, even cut.) Press straight down on the cookie cutter, using even pressure; gently press the daisy shape out of the disk. Cut the remaining daisies and place them on a clean rimmed baking sheet; cover and chill until assembly.

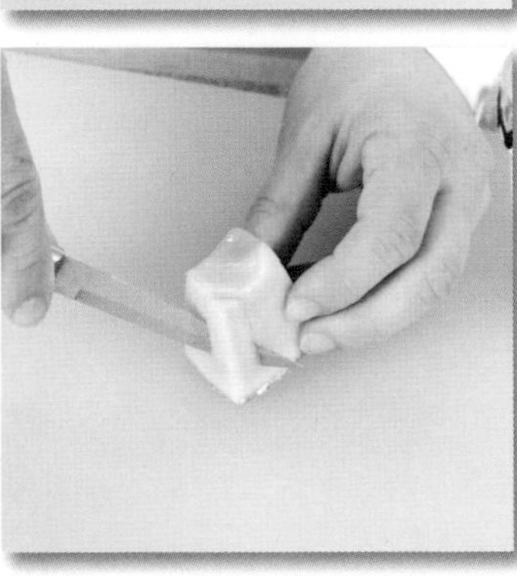

3 Slice the cantaloupe and honeydew melon into 1½″ wedges. From 1 cantaloupe wedge, cut 6 balls using a melon baller. Set aside balls for use on daisies. From 2 remaining cantaloupe wedges, cut 6 (1½ x 1½″) base chunks, leaving rind attached. Trim the 4 corners from each base piece; set into the pots, rind down. Trim any remaining flesh from wedges for later use.

4 Cut and discard the rind from the remaining wedges of cantaloupe and all honeydew. Julienne all melon flesh (including trimmed pieces from step 3) into approximately 1 x ¼″ slivers; lightly toss to combine. Gently tuck a portion of the melon slivers between the base pieces of cantaloupe and the votive cup's edges.

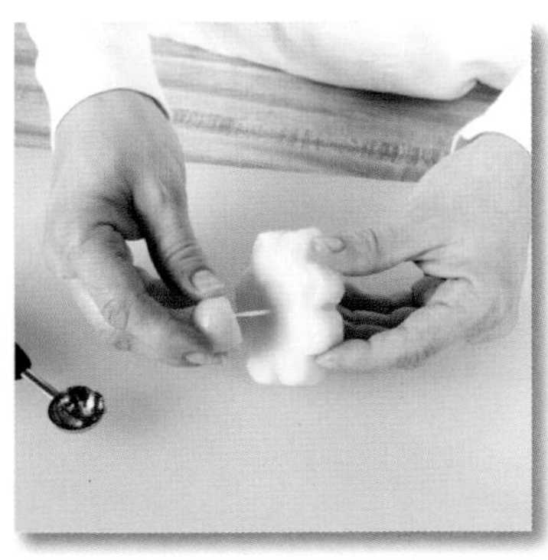

5 Break 3 toothpicks in half. Poke the blunt end of a toothpick half into the center of the pineapple flower until a ½″ point remains visible. Attach a melon ball. Repeat with remaining flowers.

6 Insert a stick into the edge of a pineapple flower, between 2 petals, sliding the pineapple down onto stick until it hits the core.

7 Insert each flower stem into cantaloupe base to stand upright. Use remaining melon slivers to fill each cup.

8 Drain the soaking apples. Insert a toothpick into the end of each apple leaf at a slight angle. Insert pick end into fruit base, positioning the apple leaf so it appears to be attached to the flower stem.

Tip: Purchase an additional apple to practice carving skills before creating the final product. Carving the leaves is not as difficult as it may appear and just a little practice should have you impressing guests soon.

Celebration Suggestions...

Bridal Shower, Easter Brunch, Garden Party, May Day Treats or Birthday Favors

Star Spangled Berries

You will need:

- **60 medium strawberries**
- **Toothpicks**
- **12 to 16 oz. white vanilla flavored candy wafers (or white chocolate)**
- **Styrofoam sheet topped with waxed paper**
- **4 oz. blue vanilla flavored candy wafers**
- **4 oz. red vanilla flavored candy wafers**
- **2 small zippered plastic bags**
- **Styrofoam (for base)**
- **Container (Sample uses a painted clay pot, 5¾″ deep and 6″ in diameter.)**
- **1 (⅝″ or ⅞″) wooden dowel, 18″ long**
- **2 cabbages, approx. 4″ and 5″ in diameter**
- **½ bunch green kale (to cover dowel and base)**
- **Green florist wire (or green twist-ties)**
- **½ bunch leafy lettuce (to cover cabbages)**
- **5 to 6 oz. blackberries**
- **4 to 5 oz. blueberries**
- **Decorative ribbon, optional**
- ***Optional:*** *Star candy mold, candy sticks*

To Begin...

1 Slide one toothpick partway into the stem end of approximately 30 cleaned, pat-dried strawberries; leave as much pick exposed as possible. In a microwave-safe dish, melt white candy in the microwave following package instructions. Spoon melted candy over each berry to cover evenly. Stand each berry upright to dry for approximately 30 minutes by inserting the end of its toothpick into waxed paper-covered Styrofoam.

2 Melt blue candy in the microwave in same manner as before. Spoon warm melted candy into one corner of a small zippered plastic bag. Push mixture down into the corner, smooth out most of the air, seal and twist the bag above the candy. With scissors, snip off a very small piece of the corner of the bag to create a "tip" for "piping." Gently squeeze melted blue candy back and forth across 20 white coated berries. Return

continued

each berry to the foam sheet to dry for 30 more minutes. Melt red candy in the same manner, and drizzle across remaining white coated berries and about half of the dried blue ones. This will result in 10 blue, 10 red, and 10 red and blue drizzled strawberries.

3 Trim the foam base to fill the pot and fit very snugly inside it. Insert the dowel, pressing it firmly into the center of the foam base.

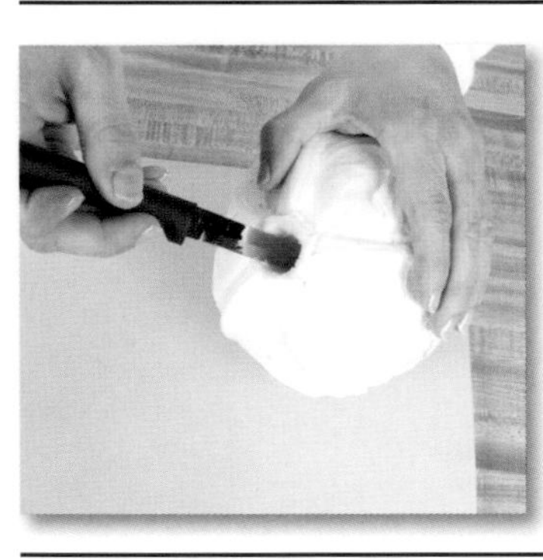

4 Remove loose outer leaves of cabbages. With a knife, cut out the core of the larger cabbage, to a depth of at least 2½″ and just wide enough for the dowel to fit inside; this is the starter hole.

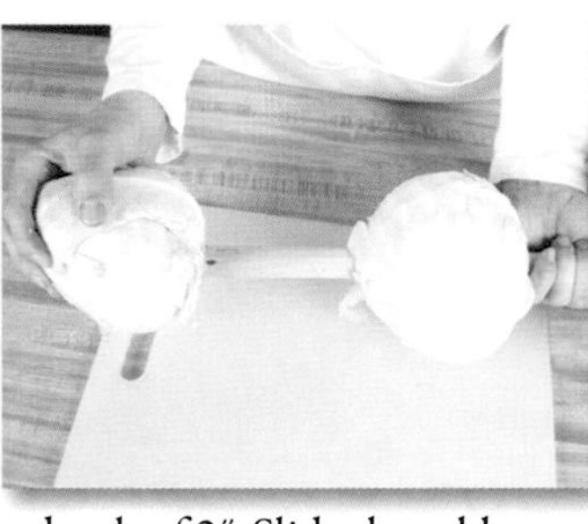

5 Remove dowel from base and push dowel into starter hole and all the way through the cabbage and slide the cabbage halfway down the dowel. Cut out the core of the smaller cabbage to a depth of 2″. Slide the cabbage onto dowel until it rests in a stable position on the dowel end. Reinsert the dowel into the base and adjust cabbage position as necessary.

6 Wrap kale around the visible portions of dowel and secure with small pieces of green florist wire.

7 Cover cabbages with leafy lettuce leaves and use toothpicks to secure them, pushing picks through the lettuce and halfway into the cabbages. Arrange the uncoated strawberries on the cabbages by sliding them stem end first onto the exposed toothpicks. Add more picks as needed to attach additional uncoated strawberries, leaving space for the coated berries to be inserted between them.

8 To fill the topiary with coated strawberries and stems of blueberries/blackberries, insert empty toothpicks between the uncoated berries already arranged. Slide coated berries, stem first, onto picks, arranging for balanced color distribution of the coated and uncoated berries. Then, slide 3 to 4 blueberries or 2 to 3 blackberries onto inserted picks, again arranging for balanced color.

---

9 Tuck leafy kale into the top of the flower pot to cover the foam base. If desired, tie decorative ribbon around the rim of the pot and drape to the side. Discard kale and leafy lettuce after fruit has been eaten.

Variation: *Molded white candy stars make an attractive addition to the topiary. Create stars by filling molds with additional melted candy wafers. Insert candy sticks for the larger size and toothpicks for the miniature size. When set, insert small star picks into the cabbage at the top of the topiary, or larger ones into the foam base.*

Celebration Suggestions...

4th of July, Memorial Day, Veterans Day, Military Homecoming Party, Political Event, or change the colors for other celebrations, such as pinks for a baby shower or red and green at Christmastime.

Doubly Delicious

You will need:

- 2 medium, round seedless watermelons
- Toothpicks
- Paper (for making a pattern)
- 2″ heart cookie cutter
- 1 honeydew melon, halved and seeded
- 1 cantaloupe, halved and seeded
- 24 to 32 (7″ to 8″) decorative plastic drink stirrers

To Begin...

1 If the melon does not sit level on a flat surface, trim a scant layer from the bottom side. Mark a line for removing the top of watermelon by inserting toothpicks around the upper perimeter. Select a line far enough below the top to allow for a wide opening, while leaving a large surface area for face carving. Slice around the melon with knife blade inserted at a slight angle; remove the top piece and retain for use in step 4.

2 Cut melon balls from the top 3″ of flesh inside the melon; set aside. To maintain a solid flat surface of flesh inside the melon (which will be used to secure fruit skewers), scrape out loose flesh. Soak up excess juice with paper towels.

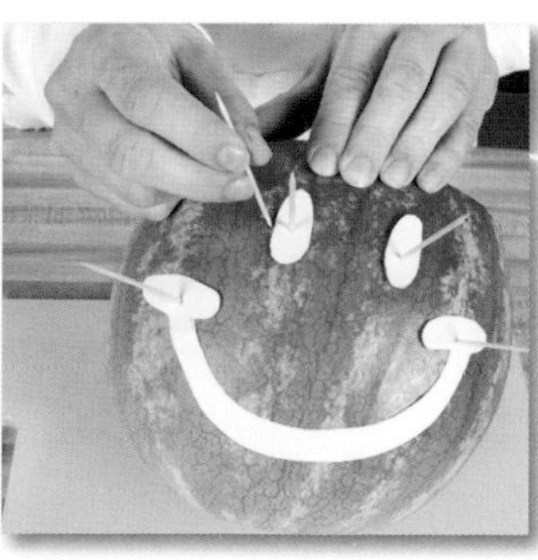

3 Make a paper pattern for the eyes and smile using the diagrams on p. 31 as a guide; adjust sizes as appropriate. Attach the patterns to the melon with toothpicks. Poke holes into the melon rind with a toothpick

continued

around the perimeter of the patterns to mark the shapes; remove the pattern. Make a very shallow cut into the rind of the melon around the mouth, using the toothpick holes as a guide. Use a small metal spoon or carving tool to remove only the surface layer of rind. Cut through the rind to remove the eyes. Repeat steps 1 through 3 to make a second melon head.

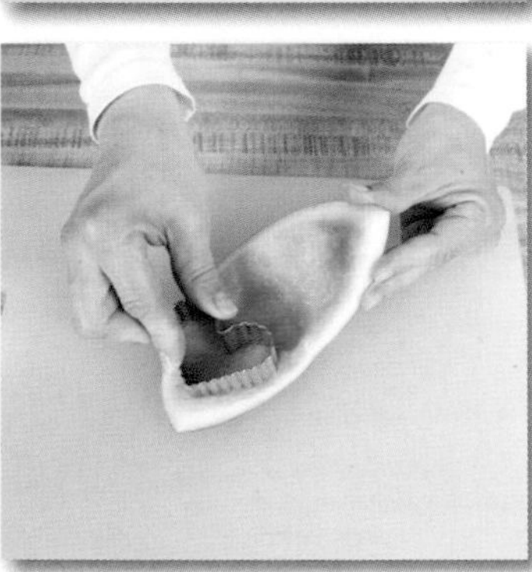

4 Trim just enough flesh from the removed watermelon tops to leave a layer of pink for the bow tie and/or heart. Make a bow tie pattern from paper using diagram on p. 31. Attach the pattern to the green side of the rind and use a toothpick to poke holes around it, marking the pattern. Remove pattern and use a small knife to cut out the tie. To cut a heart, press the cookie cutter through a removed top piece of rind.

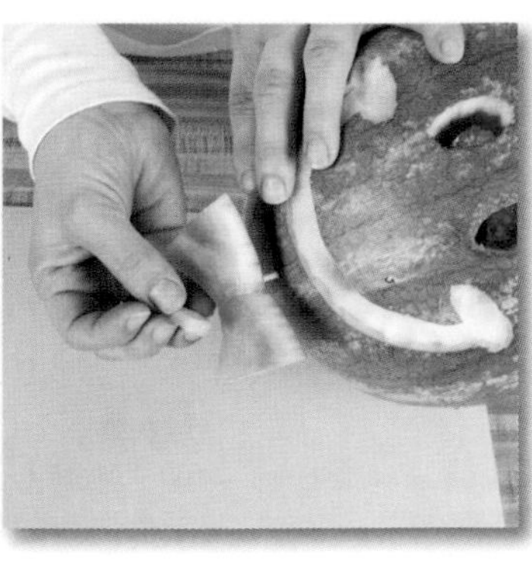

5 Insert a toothpick into the head in desired position for the tie and heart; slide tie and heart pieces onto the picks, followed by a small trimmed piece of honeydew or cantaloupe.

6 Scoop melon balls from the honeydew and cantaloupe; discard rinds. Skewer balls of watermelon, honeydew and cantaloupe onto drink stirrers, leaving approximately ½″ showing at one end and 1″ to 1½″ at the other end. Insert the longer end into the solid flesh inside the watermelon so the skewer stands upright. Repeat to fill each melon with 12 to 16 fruit skewers.

Diagrams

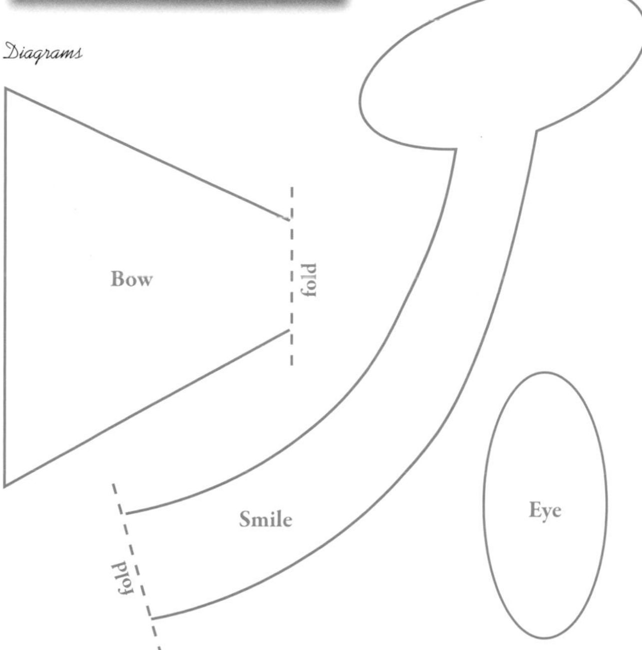

Celebration Suggestions...

Anniversary Party, Bridal Shower, Birthday (boy or girl), Farewell or say, "Welcome to the neighborhood, from both of us."

Sangria Sunset

You will need:

- 2½ to 3 lbs. red seedless grapes
- Glass pitcher (Sample is 10″ tall with a 5½″ diameter base.)
- Meringue powder
- Sugar
- 2 large oranges
- 1 lemon
- 1 lime
- 13 (12″) wooden or bamboo skewers
- Assortment of fresh fruit pieces such as kiwi, cherries, pineapple, peaches, watermelon, cantaloupe, apples*, etc.

To Begin...

1 Choose one attractive, large bunch of grapes to use as the sugared cluster that will hang from the pitcher's rim. Create an egg white wash from the meringue powder, according to label instructions. Thoroughly brush the grape cluster and a few loose grapes with the mixture. Sprinkle grapes with sugar until evenly coated. Set aside to dry for at least 2 hours.

2 Fill the bottom 2″ to 3″ of the pitcher with loose grapes. Set one whole orange into the pitcher, on top of the grapes.

3 Cut the lemon, lime and remaining orange into ¼″ to ⅜″ slices. Slide citrus slices into pitcher around outside edges, placing the slices upright. Use a butter knife to tuck grapes between the whole orange and the citrus slices to force the slices to lie flat against the glass. Add small grape clusters to the center of the pitcher (on top of the whole orange) and continue to tuck citrus slices and wedged grapes to desired height.

4 With scissors, trim remaining grapes into petite clusters (3 to 5 grapes). Tuck clusters into the container until pitcher is almost full. To make fruit spikes, slide assorted fresh fruit pieces onto the top 4″ to 5″ of 12 skewers (about 5 pieces of fruit per skewer). Slide skewers into the pitcher, poking the ends into the whole orange as an anchor. Add a few loose grapes to pitcher to fill in the remaining space around the skewers. (At this point, the arrangement can be loosely covered and refrigerated for several hours prior to service, if required.)

5 Near the handle, slide the remaining empty skewer into the pitcher, just behind the citrus fruit. Before serving, carefully drape the sugared grape cluster around the skewer so it hangs from the rim of the pitcher. Slide loose sugared grapes onto the top of the skewer to cover it. If desired, serve with Marshmallow Fruit Dip.

**To prevent cut apples from browning, see tip on page 58.*

Celebration Suggestions...

Engagement Party, Cocktail Party, 4th of July, Retirement or say, "It's 5 o'clock somewhere," "Let's fiesta!" or "Livin' la vida loca."

Marshmallow Fruit Dip

1 (8 oz.) pkg. cream cheese, softened
1 (7 oz.) jar marshmallow créme
½ tsp. vanilla extract

Beat cream cheese with an electric mixer until smooth. Add marshmallow créme and vanilla, and beat until fluffy.

Delightfully Sunny Blooms

You will need:

- Container (Sample uses a ceramic pot, 6″ deep and 8″ in diameter.)
- 1 head iceberg lettuce
- 1 bunch green kale or leafy lettuce
- 1 whole fresh pineapple
- ½ small seedless watermelon
- 2 fresh limes
- Metal cookie cutters (2½″ tulip, 3¼″ flower and 1½″ star)
- 20 to 24 large strawberries
- 55 to 60 (10″ and 12″) wooden or bamboo skewers
- 6 to 8 oz. yellow vanilla flavored candy wafers (1¼ cups)
- Styrofoam sheet topped with waxed paper
- 2 to 3 oz. white vanilla flavored candy wafers (½ cup)
- Plastic piping bag with small round tip
- 1 small bunch green seedless grapes
- ½ pt. fresh blueberries
- ½ pt. fresh blackberries
- ½ lb. fresh bing cherries with stems (or large red seedless grapes)

To Begin...

1 Place iceberg lettuce into container, trimming it if necessary to achieve a snug fit. Arrange leafy lettuce around edges as shown; set aside.

2 Slice the top off the pineapple. Cut crosswise to make 5 (¾″ to 1″) pineapple disks. Center the flower cookie cutter over one pineapple disk. (Metal cookie cutters are recommended for a clean, even cut.) Press straight down on the cookie cutter, using even pressure. Gently slide the flower shape out of the disk. Repeat to make 4 more flowers; set aside. Discard skins.

3 Cut the watermelon crosswise into several ¾″ slices. Use the tulip cookie cutter to cut 3 or more tulips from the watermelon slices, pressing the cutter gently through the fruit. Use the small star cookie cutter to cut 1 watermelon star for each pineapple flower. Set watermelon tulips and stars aside on paper towels.

Cut each lime in half lengthwise. Then cut each half into 3 wedges; set aside.

5 Poke the pointed end of a skewer into the stem end of each strawberry, stopping before the skewer pierces through the tip of the berry. Use a variety of skewer lengths. In a microwave-safe measuring cup, melt the yellow candy wafers as directed on the package. Stir until smooth. Holding a skewered strawberry over the measuring cup, spoon melted candy over berry until well coated. Tap skewer on the side of cup to let excess coating drip off. Stand berry upright to dry by inserting the end of the skewer into waxed paper-covered Styrofoam. Repeat to make 9 or 10 coated berries. When yellow coating is dry, melt the white candy wafers in a microwave-safe bowl, following package instructions. Spoon melted white coating into the piping bag. Holding a coated berry by the skewer, pipe white swirls over yellow-coated strawberries. Press skewer ends back into Styrofoam to dry.

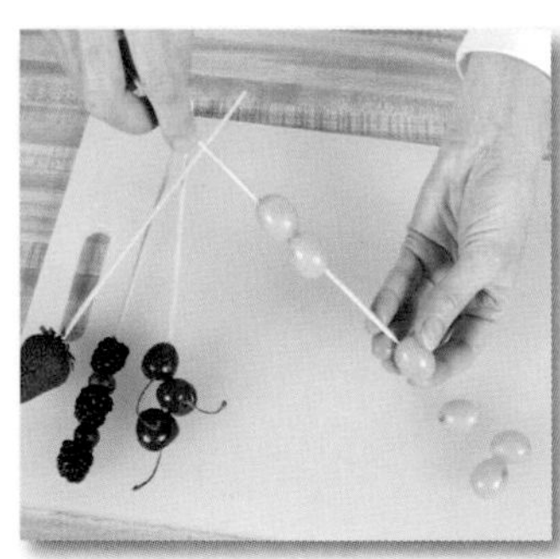

6 Use a variety of skewer lengths to create spears of cherries, green grapes and blueberries alternating with blackberries. Gently slide the pointed end of a skewer through the fruits, being careful not to pierce the top of the last piece. Make 6 to 8 cherry spears using 2 to 5 cherries on each skewer and placing them at different angles with the stems up. Make 7 or 8 grape skewers using 3 to 5 grapes on each skewer. Make 11 or 12 blueberry/blackberry skewers using 4 to 7 berries on each skewer.

7 To assemble flowers, break several toothpicks in half. Press the broken end of a toothpick into the center of each pineapple flower; fasten a watermelon star on each toothpick tip for the flower's center. Slide 1 or 2 prepared lime wedges ("leaves") on the pointed end of 8 skewers. Press a pineapple flower or watermelon tulip of top of each "leafed" skewer stem. Move lime leaves as desired.

---

8 To arrange the bouquet, push stems firmly into lettuce base until pieces are straight and secure. Insert tallest stems in the back and work toward the front where the shortest pieces will be. Cut stems with pruning shears as needed to make a variety of heights. Some stems will be pushed deeper into the lettuce base than others. Arrange the pineapple flowers and tulips first, followed by both types of strawberries. Insert remaining skewers of cherries, grapes and blueberries/blackberries to fill empty spaces and provide a balance of colors. Adjust leafy lettuce around base as needed.

Tips:

- To achieve a well-proportioned bouquet, insert 12″ skewers into these items: 1 pineapple flower, 2 watermelon tulips, 3 or 4 yellow strawberries, plus several red strawberries and fruit spears. Use 10″ skewers for remaining stems, trimming as needed.
- To make it easier to push skewers into lettuce, remove the top piece of fruit on a skewer and push down to desired height; then replace fruit.

Celebration Suggestions...

Mother's Day, May Day, Birthday, Anniversary, Valentine's Day, Congratulations, Get-Well, Thank You or say, "You're a Star!" or "I'm berry fond of you."

Bloomin'

Bloomin' Berries

Berries

You will need:

- 6 oz. cream cheese, softened
- 1 tsp. clear vanilla extract
- ⅔ C. heavy whipping cream, well chilled
- 6 T. sugar
- 30 to 40 medium strawberries
- 1 head iceberg lettuce
- Toothpicks
- Plastic piping bag fitted with small star tip
- Container (Sample uses a white pedestal bowl, 3″ deep and 6½″ in diameter.)

To Begin...

1 Using an electric mixer, beat cream cheese and vanilla until smooth; set aside. Using a well-chilled mixing bowl and beaters, whip cream and sugar until very stiff peaks form. Gently combine cream cheese and whipped cream, until just blended. Chill for 1 hour.

2 Trim the head of lettuce to fit snugly into the container, allowing the top curve of the lettuce to form a mound in the center that rises 1½″ to 2″ above the rim.

3 Set cleaned and dried strawberries, stems down, on a cutting board. Slice an "X" in the tip of each berry, cutting about ¾ of the way to the stem. Gently pry the strawberries open with fingers.

4 Insert a toothpick into the center of the top of the lettuce until approximately 1″ remains visible. Gently slide one berry onto the toothpick through its stem. Continue this process, working in a circular pattern to fill the bowl. Be sure berries are opened wide enough to fill with cream.

5 Spoon chilled cream mixture into the decorating bag. Pipe cream into pried open strawberry centers, working from the middle of the arrangement outward. Refrigerate uncovered for up to 3 hours before serving. Provide a serving fork to slide under berry stems and gently remove the berries from picks.

Variation: *This arrangement is also attractive without the cream filling. Insert toothpicks deeply into the lettuce to make them less visible in an arrangement without cream. It is a healthier option, yet still makes a beautiful centerpiece or gift.*

Celebration Suggestions...

Bridal Shower, Mother's Day, Valentine's Day; use a blue base or add blue sugar sprinkles to say, "Happy 4th of July!"

Beautiful Bounty

Beautiful Bounty

You will need:

- 2 bunches green onions
- 9 to 12 radishes
- 3 large thick carrots
- 1 large cucumber
- 4 green zucchini, divided
- 1 yellow straight neck summer squash
- 1 stalk broccoli
- ½ head cauliflower
- 1 bunch celery with leafy center stalks
- 1 (15 oz.) can whole baby corn
- 1 pt. cherry or grape tomatoes
- 10 to 16 sugar snap peas
- 1 to 2 bell peppers, color of choice (Sample uses yellow.)
- Large ceramic pumpkin container (Sample is 6″ deep and 8½″ in diameter.)
- 2 to 3 heads iceberg lettuce
- 1 bunch leafy lettuce
- 40 to 50 (10″) wooden or bamboo skewers
- 40 to 50 (12″) wooden or bamboo skewers
- Toothpicks

To Begin...

1 ***Make green onion frills:*** Trim off just the tips of the green stems and the roots of the white bulb; discard trimmings. Cut the white bulb from the green tops so that the bulb end is approximately 4″ in length; chill and reserve the green tops for later use. Hold the white bulb and make multiple slices from the center of the white portion up through the tips of the green ends. Place sliced onions in iced water for several hours to make them "frill."

2 ***Make radish roses:*** Trim the ends of the radishes. Set a radish on end on a cutting board. Make 4 vertical cuts around the outside of the radish to form a "square" at the center; the cuts should be about ½ to ⅔ of the way into the radish. Make 4 more cuts around the outside of the radish beginning at the corner of the "square" and working evenly across the other 3 corners to complete 8 cuts to the same depth. Place the cut radish in iced water to soak, making it fan out. Repeat with remaining radishes.

3 ***Make carrot flowers:*** Peel and trim ends from a large carrot. Angling the knife towards the tip of the carrot, make a shallow slice on 1 side, beginning about 1½″ to 2″ from the tip and stopping just short of the tip. Turn the carrot and slice again in the same manner. After working your way around the carrot to slice the petals, break the "petaled" tip from the carrot, giving you the hollowed flower piece and leaving the beginning tip on the main carrot to start the next flower. (You may need to poke a knife into the hollow of the flower to pop it off, being careful not to break off the petals.) Repeat to obtain 4 to 6 flowers per carrot, depending on the carrot size. Place carrot flowers in iced water to soak.

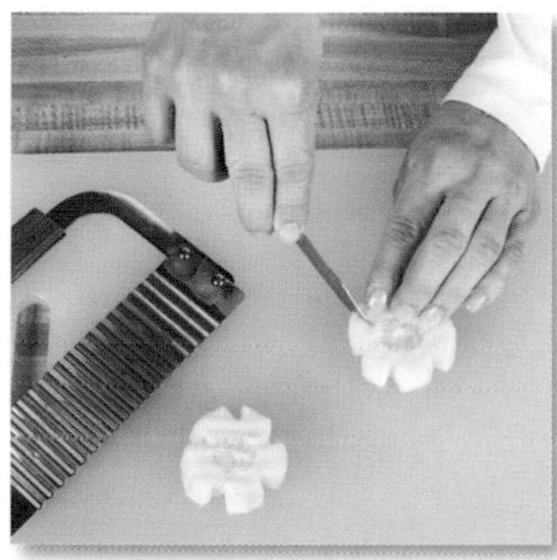

4 ***Make cucumber flowers:*** Cut peeled cucumber into ½″ slices (with a crinkle cutter, if preferred). Use a small paring knife to cut away 5 small triangular pieces from each slice, to create the petals of cucumber flowers. Lay pieces on a plate, cover with moist paper towels and plastic wrap; chill until assembly.

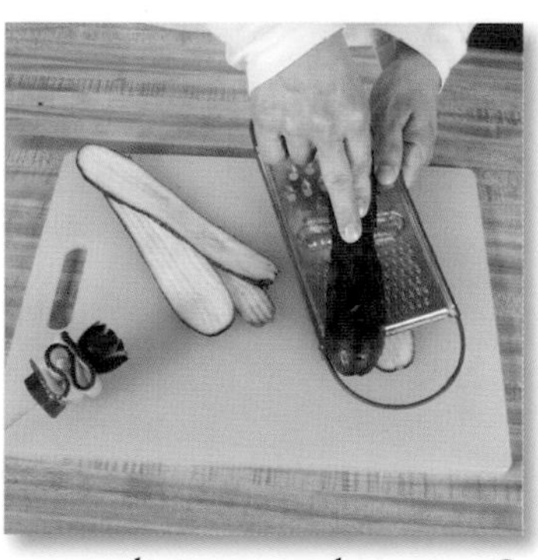

5 Use 2 zucchini to create long flat strips of zucchini to be skewered with the radish roses. Slide the zucchini lengthwise against a slicing blade, such as the one found on a cheese grater. Discard first piece that is mainly skin, then repeat slicing to make strips. Stack strips together, wrap in plastic wrap, and chill until assembly.

6 Slice the remaining zucchini and yellow squash into ⅜″ rounds. Make about 5 skewers of 3 rounds by sliding a yellow piece, then a green piece and then yellow again onto the pointed tip end of each skewer as shown. Lay skewers, as well as extra zucchini rounds, on a moist paper towel-lined tray, cover with moist towels and chill until assembly.

7 Separate the broccoli and cauliflower into florets; discard stems. Cut and discard the bottom from the celery bunch. Cut the outside stalks into long thin sticks and leave the leafy parts on the center stalks. Drain and rinse baby corn. Rinse tomatoes and peas. Slice cleaned peppers into the shapes of leaves. Place items on skewers as desired, such as stacking 3 tomatoes on a single skewer. Use combinations of 10″ and 12″ skewers, as well as toothpicks for longer items, such as celery.

8 Prepare the base by filling the container with lettuce heads, trimming them into chunks if necessary. Place a large rounded piece of lettuce in the top of the container so that it rises above the edge by several inches. Tuck leafy lettuce around the edges.

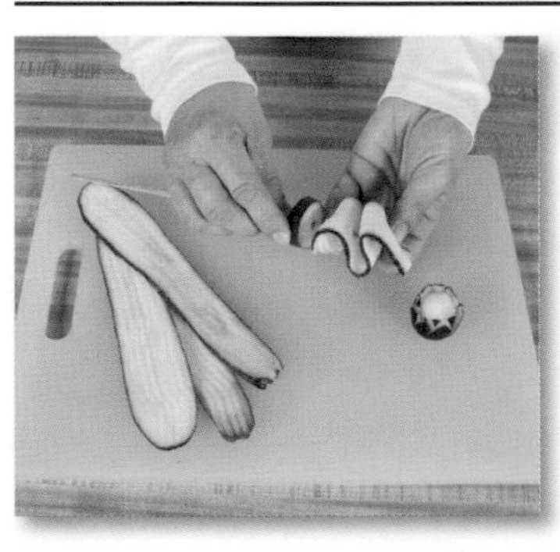

9 Make radish rose skewers by sliding a zucchini round onto the pointed end of a skewer, followed by a fan-fold zucchini strip and a radish rose. Insert the radish rose skewers into lettuce base, placing them in an even pattern.

10 Slide reserved, chilled onion greens onto skewers like sleeves; top them with carrot flowers. Slide sleeves of onion greens onto other vegetable skewers as desired.

11 Assemble the arrangement in the following suggested order: skewers of broccoli and cauliflower, celery pieces, rounds of zucchini and squash, pea pods, baby corn, 3 tomatoes, onion frills and sweet peppers. Use a combination of 10″ and 12″ skewers and trim skewers as needed for variety and balance. Serve with Golden Veggie Dip, if desired. This bouquet holds well when refrigerated for several hours; spritz with water just before serving.

Variation: *To achieve a more natured or rustic appearance for your arrangement, use a fresh, hollowed-out pumpkin as the base of your vegetable bouquet.*

Celebration Suggestions...

Thanksgiving, Fall Festival, Halloween or use a holiday container to say "Merry Christmas." Use a container displaying the colors of a favorite sports team to say, "Go Team!" or with company logo for a Retirement Party Bouquet.

Golden Veggie Dip

1 C. mayonnaise
1 tsp. ground turmeric
1 tsp. garlic salt
2 tsp. sugar
2 tsp. dried minced onion
1 tsp. prepared horseradish
1 tsp. white vinegar

Stir together mayonnaise, turmeric, garlic salt, sugar, onion, horseradish and vinegar in a small serving bowl. Chill for at least 2 hours prior to serving.

Caramel Apple Gourmet Bouquet

You will need:

- Styrofoam (for base)
- Container (Sample uses a 6″ square tin box, 6″ deep.)
- Chocolate sprinkles
- Red cinnamon flavored sprinkles
- 4 (1.4 oz.) toffee candy bars, chopped
- 1 (3.1 oz.) box Sno-Caps (semi-sweet chocolate nonpareils)
- 16 oz. dry roasted peanuts (Chop 1 cup.)
- 13 medium crisp, tart apples
- 13 wood craft sticks
- Toothpicks
- 1 (12 oz.) bag pretzel rods
- 12 to 14 (10″) thin wooden or bamboo skewers
- 13 (12″) heavy wooden or bamboo skewers
- 10 C. caramel bits* (about five 11 oz. bags), divided
- ½ C. heavy whipping cream, divided
- Styrofoam sheet topped with waxed paper
- 8 oz. vanilla candy coating or almond bark, melted
- 8 oz. chocolate candy coating or almond bark, melted

To Begin...

1 Trim foam to fit snugly into container, so that the top of the foam will be 2″ below the top of container. Wrap foam in foil and place into container.

2 Prepare individual plates of candy and nut toppings. Wash and dry apples well to remove any wax coating. Insert one craft stick into the stem end of each apple.

3 Gently twist a toothpick into the end of each pretzel rod, creating a starter hole. Remove toothpick and gently insert a 10″ thin skewer approximately 1″ deep, into the pretzel

continued

end, twisting carefully. (If pretzel tip breaks, remove jagged ends, make new starter hole and skewer into the white/broken end of the pretzel.

4 Combine 7½ cups caramel bits with ¼ cup + 2 tablespoons cream in a microwave-safe container. Microwave, stirring every minute, for 4 minutes or until very smooth. Add up to 2 tablespoons of water to thin if necessary. Holding each apple by the stick, dip each apple into the caramel mixture; let excess drip off and scrape gently against a spoon to smooth caramel. Hold and twist momentarily to create a smooth finish and stop drips. To create a scalloped edge, tilt apple into caramel with each turn of the stick. (Reheat caramel if necessary; combine any remaining caramel mixture with new caramel mixture in step 5, to coat pretzels or additional apples.) Finish as follows:

Chopped peanuts, chocolate and red sprinkles – Dip and roll top of freshly coated caramel apple in topping, place upright to set by inserting stick into waxed paper-covered Styrofoam sheet.

White chocolate with toffee – Place scalloped-edge caramel apple upwright into covered Styrofoam; once set, tilt apple into melted vanilla candy, creating a white scalloped edge. Then immediately dip coated end of apple into toffee pieces and dry upright in foam sheet.

Chocolate and vanilla drizzle – Insert freshly coated caramel apple in foam sheet until set; drizzle melted vanilla candy over the top of the apple, followed by melted chocolate, then dry upright in foam sheet.

Nonpareils – Insert freshly coated caramel apple in foam sheet until cooled, but not quite set; gently press nonpareils into caramel and return to foam sheet to finish setting.

5 Prepare more caramel mixture using remaining ingredients. Holding each pretzel, spoon caramel over half of rod to coat. Finish by topping pretzels using methods described in step 4.

6 Poke a heavy skewer into each apple near the previously inserted craft stick. To arrange, insert apple skewers into the base, starting with the bottom row angled outward and working toward the top/center of the bouquet; shorten skewers as necessary. (Sample has an 8 apple bottom layer, 4 apple middle layer and 1 apple top.) Remove one apple skewer while filling top of container with peanuts. To add pretzels to bouquet, shorten skewers as necessary, remove pretzels from skewers while inserting them and slide pretzels back onto placed skewers.

**Find caramel bits with the baking supplies in most supermarkets. Similar in size to chocolate chips, bits come unwrapped for easy melting. If unavailable, use 4 (14 oz.) bags of wrapped caramels in their place.*

Tip: Caramel apples and pretzels maintain their freshness well, without refrigeration.

Celebration Suggestions...

Fall Festival, Oktoberfest, Thanksgiving, Teacher Appreciation, or create an arrangement with a casual Halloween base to say, "Trick or Treat!"

Trees to Relish

For the large tree, you will need:

- **5 x 18″ Styrofoam cone**
- **2 bunches green kale**
- **1 large head fresh cauliflower**
- **1 to 2 bunches fresh broccoli**
- **Toothpicks**
- **1 bag carrot chips**
- **4″ cellophane frill toothpicks**
- **20 to 25 large blue cheese-stuffed olives**
- **1 pt. cherry tomatoes (20 to 25 tomatoes)**
- **2 yellow and/or red sweet bell peppers, halved and seeded**
- **2½″ star cookie cutter, optional**

For the medium tree, you will need:

- **4 x 12″ Styrofoam cone**
- **1 large bunch purple leafy lettuce**
- **¼ to ½ lb. Colby Jack cheese, mild cheddar and/or Pepper Jack cheese**
- **¼ lb. summer sausage**
- **4″ cellophane frill toothpicks**
- **1½″ star cookie cutter**
- **1 (8 oz.) pkg. sliced Provolone cheese**
- **½ pkg. sliced pepperoni**
- **Toothpicks**
- **½ (6 oz.) can black pitted olives**
- **½ (5¾ oz.) jar green stuffed olives**

For the small tree, you will need:

- **4 x 9″ Styrofoam cone**
- **1 bunch green leafy lettuce**
- **Toothpicks**
- **Styrofoam sheet topped with waxed paper**
- **4 oz. white almond bark**
- **5 dried pineapple rings**
- **12 pecan halves, slivered**
- **25 large pitted dates**
- **2 to 3 oz. cream cheese, softened**
- **Plastic piping bag fitted with a long round tip**
- **1 to 2 oz. dried apple slices (24 to 26 slices)**
- **2½″ cellophane frill toothpicks**
- **5 dried pears**
- **12 dried apricots**

To Begin...

Wrap each foam cone with foil. Then proceed as directed for each tree. After covering cones with greens, they can be refrigerated until ready to decorate. Loosely wrap finished trees in plastic and refrigerate for up to 3 hours before serving, if desired. Group the trees together for one large table display, or arrange trees on separate large serving platters, woven mats or shallow baskets.

For Large Vegetable Tree:

1 Cover the large cone with kale, starting at the top of the cone and wrapping stems downward diagonally around cone. Insert toothpicks through the stems to fasten kale to cone, arranging ruffled leaves to cover foil.

2 Cut broccoli and cauliflower into medium-large florets. Poke a toothpick into the bottom of each floret. Make 30 to 32 cauliflower picks and 30 to 32 broccoli picks. With a frilled toothpick, skewer 2 crossed carrot chips together. Make approximately 20 carrot skewers.

3 Starting at the bottom of the cone, insert the carrot skewers into the "tree" in a diagonal line, which winds around the tree from bottom to top. Insert a row of cauliflower picks above the line of carrots. Insert a row of broccoli picks above the cauliflower.

4 In the space above the broccoli, insert toothpicks in a staggered pattern, leaving 1″ to 1½″ of each toothpick sticking out. Attach tomatoes and olives to the toothpicks as desired. Slice yellow bell pepper into pointed strips and attach strips to tree with frilled toothpicks, arranging them diagonally between the tomatoes and olives.

5 If desired, use the cookie cutter to cut a star from a large flat section of red or yellow bell pepper. Insert a toothpick into the top of tree and slide star on toothpick. Serve with Creamy Dill Dip, if desired (recipe on p. 57).

For Medium Meat & Cheese Tree:

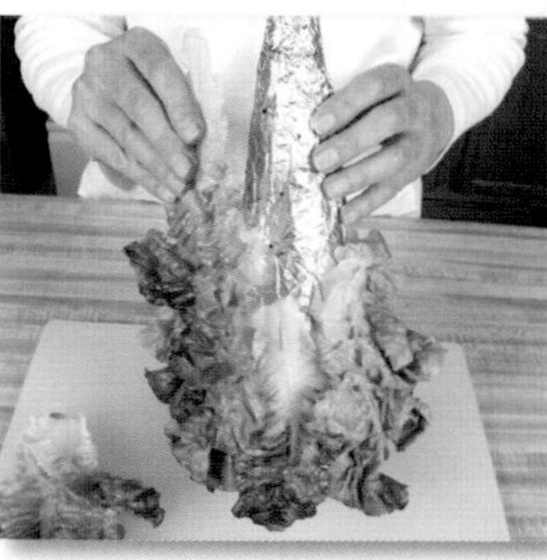

1 Cover the medium cone with purple leafy lettuce, placing the largest leaves around the bottom of cone first. Fasten the leaves to the cone, colorful side out, by pushing toothpicks through the thick center portion of leaves. Allow leaves to hang slightly over bottom edge of cone. Work upward on the cone with successively smaller leaves, overlapping them to look like tree boughs. Cover the top of cone with the smallest inner leaves, leafy ends up.

2 To prepare the meat and cheese skewers, use a small ripple cutter or sharp knife to slice Colby Jack and/or Cheddar cheese into ½″ to 1″ cubes. Cut the summer sausage into ¾″ slices and cut each slice into 4 wedges. With frilled toothpicks, skewer a small rippled cheese cube on top of each summer sausage wedge. With the cookie cutter, cut stars out of the Provolone slices. Fold 3 round pepperoni slices in half; stack them with folded edges together and fan them slightly. Skewer 1 cheese star on top of each pepperoni fan. Skewer additional Provolone stars on top of 1″ Cheddar cheese cubes. Make other meat and cheese combinations, such as cubes of Pepper Jack on each side of a flat pepperoni slice.

3 Arrange the spears around the tree randomly, pushing the picks into the lettuce and foam just until secure. Poke plain toothpicks into the empty spaces on tree and attach black and green olives. Top the tree with several skewers of black and green olives.

For Small Dried Fruit Tree:

1 Cover the smallest cone with green leafy lettuce, using the same placement and overlapping method as for medium cone.

2 Cut each pineapple ring in half. Melt almond bark in the microwave following package instructions; stir until smooth. Poke a toothpick into each pineapple piece and spoon melted almond bark over one end of 8 pieces. Insert the toothpicks into wax paper-covered Styrofoam sheet to dry. Cut remaining pineapple pieces into small wedges

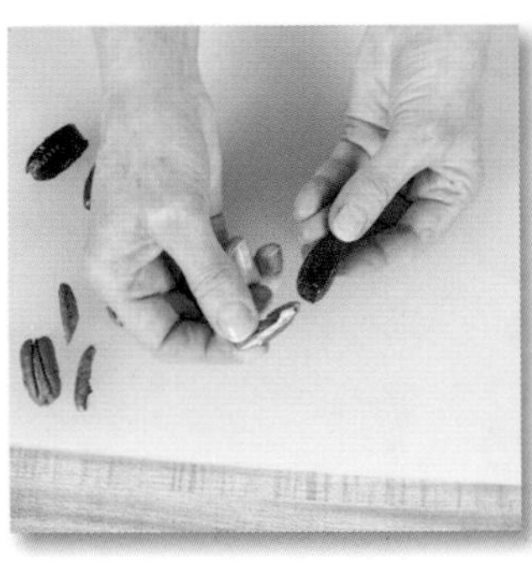

3 Insert a pecan sliver into the hole in each date. Beat cream cheese until creamy and place it into a plastic piping bag fitted with a long metal tip. Pipe a small amount of cream cheese into each end of dates.

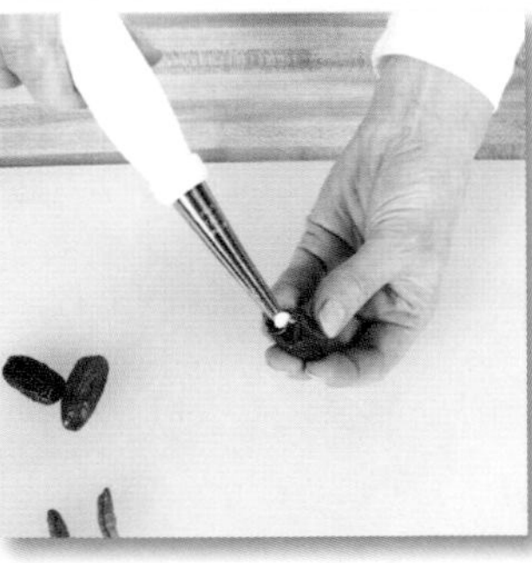

4 Fold 3 apple slices in half and stack them together, alternating cut and folded edges. Skewer the stacked apples on frilled toothpicks. Cut each dried pear in half lengthwise.

5 Remove toothpicks from dried pineapple halves. With frilled toothpicks, attach pineapple pieces around the bottom of tree, angling them as shown. Working upward on tree, use frilled toothpicks to attach a row of apricots around tree. Attach stuffed dates at an angle on the next row. Attach dried apple skewers next, followed by diagonal pear pieces and another row of diagonal dates. Add small pineapple wedges near top. Use a toothpick to attach an apricot, on edge, to the top of tree.

Variation: *Instead of using dried fruit, use fresh fruit on the small tree, or create an additional tree to serve more guests. Include dipped chocolate strawberries as a special touch.*

Celebration Suggestions...

Family Gathering, Class Reunion, Potluck Party, Office Party, or Christmas, Thanksgiving or New Year's Buffet

Creamy Dill Dip

1 C. sour cream
¾ C. mayonnaise
1 T. dried minced onion
½ tsp. seasoning salt
1½ tsp. dried dillweed
1 tsp. sugar

In a small bowl, mix all ingredients together until well blended. Cover and refrigerate for 4 or more hours to blend flavors. Transfer to a serving bowl just before serving with raw vegetables.

Tips and Special Effects

Cutting & Carving Tips

- Use a ripple potato cutter to slice melons, strawberries, carrots or other firm fruits, vegetables and cheeses. Hold food in place on a cutting board and slowly press down with the cutter to make thick or thin slices with a rippled edge. Rippled wedges of honeydew melon make attractive "leaves" in fruit bouquets.
- Add interest and texture to fruit bouquets by carving designs out of a fruit's skin. With a toothpick, poke a pattern into the skin of the fruit, then cut a shallow line into the flesh of the fruit and carefully pop out the carved pieces using the knife tip. The fruit's flesh is exposed against the contrasting skin. (See apple on page 21.)
- Some cut or carved fruits turn brown when exposed to air, such as apples. Treat them with an anti-browning solution after preparation. Several methods can be used, including brushing cut fruit surfaces with lemon juice, soaking fruits in a mixture of approximately ¼ cup lemon juice for each quart of cool water, or using a commercial product prepared as directed. Drain and pat dry before using.

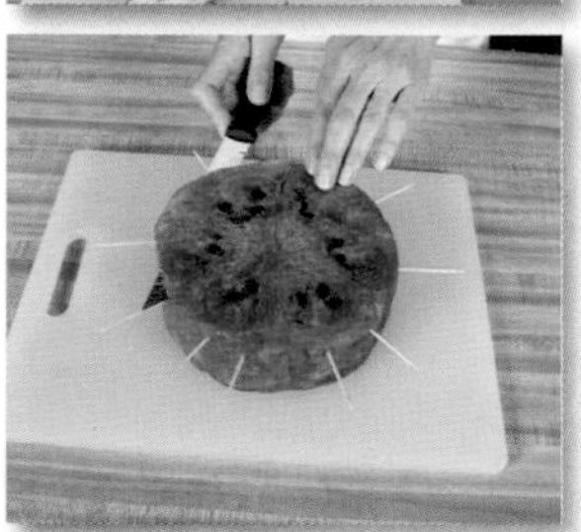

- A round pineapple or melon tends to roll when you slice it. To make cutting through the tough skin easier and safer, first cut off a small slice from one long side to make a flat edge. Place the fruit on the flat edge and then cut fruit into slices as desired.
- When the cutting needs to be very accurate, mark the fruit with toothpicks first. Remove toothpicks to expose the pattern of holes, or leave the toothpicks in place and use a sharp knife to slice the fruit along the toothpick markings.

- To serve dips, make a fun serving dish or cup from a melon, citrus fruit, bell pepper or cabbage head. Fill the edible dishes with dip just before serving. Use fruit cups with a sweet dip such as the one on page 35. Fill vegetable serving dishes with savory dips such as those found on pages 47 or 57.

Melon dish: Cut the top third or half off a melon and scoop out seeds. Cut the flesh into balls or chunks as needed for a fruit bouquet. With a large spoon, scrape out remaining flesh to leave a sturdy shell, at least ½" thick. Use a paring knife to cut the top edge into scallops, ripples or another pattern. Use a vegetable peeler to trim the melon edges.

Bell pepper bowl: Find the flattest side of the pepper; that will become the bottom of the bowl. Use a paring knife to cut out a large opening from the top side of pepper. Cut out the seeds and connective tissue inside but leave the stem in place. Trim opening edges as desired.

Citrus cup: Cut a small slice off the bottom of a large lemon, orange or grapefruit so it will sit flat. Slice off the top third or half. Use a paring knife to separate the fruit from the rind, cutting around white portion. With a large spoon, scoop out fruit and reserve it for another use, such as making lemonade. Scallop the top edges as desired. Scrape out the inside of the fruit before filling with dip.

Cabbage bowl: Cut a slice off the bottom of a head of cabbage so it sits flat. With a knife, carefully hollow out the center of the head, leaving a sturdy shell to hold the dip.

Say "Celebrate" With More Special Effects

Add candy, nuts or other treats to complement a fruit bouquet for versatility and visual interest, such as the candy stars in Star Spangled Berries on page 24 or pretzel sticks and peanuts in Caramel Apple Gourmet Bouquet on page 48. Another example would be the inclusion of wrapped candy canes in a Christmas bouquet.

Tinted melted candy coating or almond bark can be used on whole fruits, such as berries or apples, to highlight a particular holiday or to celebrate school spirit.

Use themed cookie cutters to cut shapes out of fresh pineapple and add holiday flair to a traditional edible bouquet. Cut pineapple into slices, ¾″ to 1″ wide. Press cookie cutters through the flesh and remove the shape; drain on paper towels. Melt candy coating or almond bark following package instructions, adding gel food coloring if desired. (Two ounces of bark will coat several pineapple shapes.) Insert skewers into pineapple shapes as desired. Spoon the melted bark over the top and edges of pineapple pieces, letting excess drip back into bowl. Set on waxed paper to dry, adding eyes and other details with candy sprinkles or nonpareils while still wet. When bark on front side is dry, coat the back side of each shape. Add these shapes to any general purpose fruit bouquet.

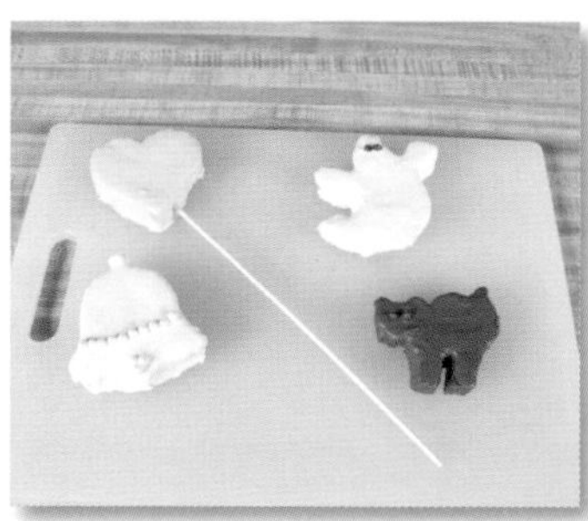

Here are some suggestions...

Halloween: Coat a cat with chocolate almond bark and/or coat several ghosts with vanilla almond bark for a Halloween bouquet. Add black candies for eyes.

Wedding or Anniversary: Coat a bell with vanilla almond bark and embellish with pearl candies.

Valentine's Day, Anniversary or Engagement: Tint vanilla almond bark with red gel food coloring and coat several pineapple hearts.

Birthday: Cut gift shapes and coat them in vanilla almond bark that has been tinted in various colors. Add details with candies or drizzle with "ribbons" of contrasting bark.

Christmas: Tint vanilla almond bark green and coat several pineapple Christmas trees, adding small colorful candies for strands of "lights."

New Baby: Coat ducks or footprints in yellow-tinted vanilla almond bark.